WHAT SHOULD BE KNOWN

ABOUT

SCIENCE LABORATORY TECHNOLOGY

AGBATI YUSUF BAMIDELE.

TABLE OF CONTENT

Chapter Two: Science Laboratory Technology as a Profession.

- Science Laboratory Technology as a Profession
- Governmental Recognition of S.L.T
- Job opportunities as a Science Laboratory Technician/Technologist
- Job entry requirements as Science Laboratory Technician/ Technologist
- Functions of Science laboratory technologist at work.
- Carrier development.
- Assessment

Chapter Three: Frequently Asked Questions

Conclusion

FOREWORD

All praises and adourations are unto the Almighty God who has been my spiritual tutor through inspiration and His noble signs on earth.

This work is done to ameliorate the misconception that has filled the heart of most Nigerian students especially those of sciences.

Studying Science Laboratory Technology in Universities or mostly Polytechnics is usually seen as a second option course since many nowadays students are jostling for the competitive courses like Medicine, Pharmacy and the likes which is usually a game. This book thereby enlightens the students on the great opportunity been overlooked.

To my fellow technicians, some are not really well informed about the studying or studied course which renders them sometime useless in their field.

This book as well contains information that will be beneficial to incumbent and prospective even the done science laboratory technology student and whole society who can find it useful.

I shall hereby acknowledge any complain and comment on this work which could be forwarded to the author's contact given below. Have a wonderful reading and knowledge acquisition.

Agbati Yusuf Bamidele.

Author (2015)

07036548720

yusbam92 @gmail.com

CHAPTER ONE

SCIENCE LABORATORY TECHNOLOGY AS A COURSE

1.1   MEANING OF SCIENCE LABORATORY TECHNOLOGY AS A COURSE

Science laboratory technology is a 'phrase' that comprises three words: science, laboratory and technology.

Let us elucidate the meaning of each word so as to have a vivid understanding of this subject of discussion.

Science which emanates from a *Latin* word *'scientia'* meaning knowledge is a systematic enterprise that builds and organizes knowledge in the form of testable explanations about the universe.

It is also deciphered to be the intellectual and practical activities encompassing the systematic study of the structures and behaviour of the physical and natural world through observation and experiment. Someone who studies science is generally called **Scientist.**

Laboratory is a room or building equipped for scientific experiments, research, and teaching or for manufacture of chemicals.

Technology is the application of scientific knowledge for practical purposes especially in industry.

It originates from Greek words *'techne'* which means art, skill and *'logia'* which means study. Technology is therefore the making, modification, usage of knowledge (science) of tools, machines, techniques, craft, systems and method of organization, in order to solve a problem, improve a pre-existing solution to a problem, achieving a goal, handle an applied input/output relation or perform a specific function.

Therefore, *Science Laboratory Technology* as a course of study is a course that impacts the skills, techniques needed in dealing with experimental activities (science) in a building set up for the purpose of the experiment (laboratory) using scientific tools, methods, systems, arts so as to achieve the predestined goal.

Science and its related fields requires laboratory for attaining optimal performance, therefore someone who will man the laboratory is needed, they are called laboratory technicians or technologist which depends on qualification.

1.2 OBJECTIVES OF SCIENCE LABORATORY TECHNOLOGY AS A COURSE

The aims of introducing Science Laboratory Technology as a course into tertiary institutions are the following amongst others.

i.   To cause rapid development of the nation through science laboratory technology.

ii. To advance SLT profession in Nigeria.

iii. To maintain high ethical standard in SLT practice.

iv. To maintain efficiently and effectively science laboratories in the context of the national science technology and innovation system.

v. To impact the skills, techniques and knowledge needed in the handling of science laboratories.

vi. To work in and/or manage laboratories of science based

industries, research institutes and institutions of higher learning.

vii. To manage quality control and product development units in industries.

viii. To teach Science Laboratory Technology subjects as instructors in Polytechnics.

ix. To repair, modify, and adapt simple instrument and apparatus in the laboratories.

1.3 LABORATORY

In a scientific investigation, there is an attitude that demands of one to explore his environment or the unknown, develop the relevant hypothesis in order to establish its acceptability or otherwise. This type of investigation can be carried out in a laboratory.

A laboratory is a setting or place to try out specific information and test principles and theories in science. It is, thus, a place or room where scientific experiments are carried out. It is appropriately equipped with various items that are required in the urge to provide answers to intriguing questions in science.

1.3.1 VARIOUS SCIENCE LABORATORIES.

Due to high diversification of science and the equipment and materials, having a general laboratory where all forms of experiment can be carried out is nearly impossible. Science laboratories are therefore classified based on the specification of the laboratory. Majorly, these are:

- Chemistry laboratory for chemistry and its related fields.
- Biology laboratory for biology and its related fields.

- Physics laboratory for physics and its related fields.

At times, any of the above could be merged purposely to achieve the aim of the laboratory which could be for research, teaching, testing etc. purposes.

The reader should bear in mind that the discussion here is strictly of non-social science.

1.3.2   SOME LABORATORIES AND THEIR SPECIFICATIONS.

- **Medical Laboratories:** For medical diagnosis and all medical experiments, test and research.

- **Pharmaceutical Laboratories:** For drug compounding, testing, researching and quality controlling.

- **Agricultural Laboratory:** For test, research on agriculture e.g. Soil analysis

- **Quality Control Laboratories:** For the confirmation of the specificity of goods or materials.

- **Electronic Laboratory**

NOTE: *All science fields require laboratory and they are called base on the specification of the laboratory.*

1.4   SCIENCE LABORATORY TECHNOLOGY AND MEDICAL LABORATORY TECHNOLOGY

Science Laboratory Technology as a course encompasses all the techniques needed in all <u>science laboratories.</u> Though, specification arises as one proceeds in the program.

Medical Laboratory Technology has a specification on the medical field. It teaches the skills required for optimum performance in medical related field tests and researches. Same does the pharmaceutical technologist.

Assessment

- What concept do you have about Science Laboratory Technology as a course before and after reading this chapter?

- In **few** sentences, discuss the meaning of Science Laboratory.

- Highlight other laboratories and their specifications. At least ten.

- What effect do you think your studying Science Laboratory Technology will have on your lifestyle?

CHAPTER TWO

S.L.T AS A PROFESSION

2.1   S. L. T AS A PROFESSION

A person who is a University graduate of science of is called *Scientist* or a specific name as perculiarized to the course. E.g Biologist, Chemist, Botanist, Zoologist etc

While those who have been trained with laboratory skills in Polytechnics are usually called *Technician* or *Technologist*.

Therefore, Science Laboratory Technology is a profession as other professions do.

2.2 GOVERNMENTAL RECOGNITION OF S.L.T

The development of Nigeria through science and technology prompts her to form an institute called National Institute of Science Laboratory Technology (NISLT) which was proposed in July 1971 by the then Federal Commissioner of Education, **Late Mr A.Y Eke.**

This proposal materialized when on 25th of March, 1972 the Nigeria institute of Technology (NIST) the forerunner of NISLT was inaugurated in Ibadan.

In 1994, the institute started the processes of transformation when it directed its effort towards obtaining statutory recognition. Toward the middle of the year 2003, the transformation processes materialized when the Act of the National Assembly No 12 of 2003 establishing the institute was gazette.

After the transformation of NIST to NISLT, the empowering of professional body an enabling act commences.

In a nutshell, SLT has govermnental recognition.

The NISLT as a national body performs the following functions:

a. Advancing Science Laboratory Technology Profession in Nigeria

b. Determining the standard of knowledge, exposure to equipment, practicals and skills, to be attained by person seeking to become registered members of the profession and reviewing those standards, from time to time, as circumstances may require

c. Promoting the highest standards of competence, practice and conduct among the members of the profession

d. Securing in accordance with the provisions of the Act, the establishment and maintenance of a register of members of the profession and the publication from time to time of the lists of those persons

e. Serving as an agency to secure, safeguard and advance the professional knowledge, standing, efficiency and interests of science laboratory technologists through the Council

f. Conducting examinations and granting certificates and diplomas and advising on, assisting in examinations relating to science laboratory technology in Nigeria

2.3 JOB OPPORTUNITIES AS A SCIENCE LABORATORY TECHNICIAN/TECHNOLOGIST

2.3.1 Job description

Scientific Laboratory Technicians are involved in a variety of laboratory-based investigations within biological, chemical, physical and life science areas.

They may carry out sampling, testing, measuring, recording and analysing of results as part of a scientific team. Technicians provide all the required technical support to enable the laboratory to function effectively, while adhering to correct procedures and health and safety guidelines.

Scientific laboratory technicians carry out work that assists in the advancement and development of modern medicine and science. The work plays an important role in the foundation stages of research and development (R&D) and in scientific analysis and investigation.

They are mainly employed within industry, in government departments and research organisations.

The role of a teaching laboratory technician is similar although their work takes place in educational institutions, where they support science teachers, lecturers and students

2.3.2 Employers and Vacancies Sources

Many public and private organisations employ scientific laboratory technicians. These include:

- large public limited companies in industry;
- hospitals and public health organisations;
- specific government departments and agencies or government-funded research institutions;
- environmental agencies;
- utility companies;

- research and forensic science institutions;
- Pharmaceutical and chemical companies.

There are many companies in the food manufacturing business where technicians could seek employment. There are also a range of companies involved in the manufacture of:

- plastic;
- metal;
- oil;
- cosmetics;
- food and textiles etc.

The education sector is a large employer of laboratory technicians. They play great role in schools, universities and educational research centres.

2.4 Job Entry Requirements as Science Laboratory Technician/Technologist

It is not essential to have a degree to become a scientific laboratory technician as many posts ask for GCSEs or science-related A-levels (or equivalent).

However, many technicians do have degrees and so holding a higher qualification in a relevant subject can be useful for securing a job, particularly if competition is high.

A HND or degree in one of the following subjects could be helpful:

➢ biology;

➢ biomedical science;

➢ biotechnology;

➢ chemistry;

➢ environmental science;

➢ forensic science;

➢ materials science/technology;

➢ pharmacology;

➢ physics.

Any degree that has a technical, Industrial Training or scientific element will be useful. A Pre-entry Postgraduate qualification is not required. Employers value Pre-entry experience in a laboratory, as it not only demonstrates your familiarity with lab procedures, but also shows your commitment and interest in the field. If your degree does not include a year in industry, try to gain some part-time or voluntary work in a laboratory or scientific setting. You could approach employers to see if it would be possible to work-shadow someone in their company.

For Nigerian Polytechnics, one year Industrial Training program could be helpful.

You need to show evidence of possessing the following **skills** as they are expected from you:

- the ability to learn specific, practical techniques and apply this knowledge to solve technical problems;
- good hand and eye coordination and the ability to use technical equipment with accuracy;
- the ability to maintain and calibrate technical equipment;
- time management skills in order to work on several different projects at the same time;
- flexibility in order to work with and provide support for a number of people;

- excellent oral communication skills in order to work effectively with colleagues from all parts of the organisation and to explain complex techniques to interested parties;
- experience in providing demonstrations and writing technical reports;
- teamwork skills and patience;
- attention to detail.

Excellent record-keeping skills are required, along with basic maths and computing. As you progress through your career, you may also need to learn management and leadership skills.

Any previous work experience, even if it is not science-related, will be advantageous if it demonstrates that you have some of the above skills. It is helpful to stay up to date with developments in the sector and becoming a member of a professional body can help with this. Relevant organisations include:

- Institute of Physics (IOP)
- National Institute of Science Laboratory Technology (NISLT)
- Royal Society of Chemistry (RSC)
- Society of Biology

Competition varies from moderate for biological and environmental sciences, to relatively low for physical sciences. Speculative enquiries are often welcome.

2.5 Functions of Science Laboratory Technologist at Work.

Scientific Laboratory Technicians carry out the work that allows scientists to concentrate on, and perform, the more complex analytical processes in the laboratory.

Tasks can vary depending on the specific employer but typically involve:

- performing laboratory tests in order to produce reliable and precise data to support scientific investigations;

- carrying out routine tasks accurately and following strict methodologies to carry out
- analyses;
- preparing specimens and samples;
- constructing, maintaining and operating standard laboratory equipment, for example centrifuges, titrators, pipetting machines and pH meters;
- ensuring the laboratory is well-stocked and resourced;
- recording and sometimes interpreting results to present to senior colleagues;

- using computers and performing mathematical calculations for the preparation of graphs;
- keeping up to date with technical developments, especially those which can save time and improve reliability;
- conducting searches on identified topics relevant to the research;
- following and ensuring strict safety procedures and safety checks.

The actual nature of the work will depend upon the organisation. For example, within an environmental health department, the work may involve analysing food samples to consider prosecution and to protect public health, while within the water industry the work will mainly focus on the collection and analysis of water samples.

2.6   Carrier Development.

As a scientific laboratory technician, your career could develop through the following roles:

- assistant technician;
- technician;
- senior/lead technician;
- team leader technician;
- Laboratory manager.

As you progress you will take on more responsibility as well as supervision and management of a team of staff and the laboratory. You will carry out more complex tasks, which could include some analysis.

In order to gain promotion you may need to move to a larger employer or a role in industry wherebprogression is typically more defined. Teams are often larger and therefore provide more roles and management levels

It may be possible to become a specialist in your field. For example, in healthcare with experience and possibly further training, you could become a phlebotomist, cardiographer or physiologist.

Taking further qualifications such as a Masters or PhD and acquiring specialist knowledge may enable you to move into scientific research.

Science and research companies tend to have strong international links, which could provide the opportunity to work abroad.

Assessment

- Mention nothing less than ten areas of work as a science Laboratory Technologist.
- Briefly describe SLT as a profession.
- Highlight the expectation of a seasoned lab technician. Do you possess these expectations?

- Highlight the novel knowledge you have acquired through this chapter.
- Briefly discuss the role of NISLT in Nigeria.
- Briefly describe the possible role of Science Laboratory Technology in the nation development.

- With what you have learnt, are you ready to be a lab scientist?

CHAPTER THREE

FREQUENTLY ASKED QUESTIONS

This chapter answers some questions that are usually asked by incumbent and prospective and passed student of Science Laboratory Technology.

Q: Is it only Polytechnics that offers the studying of S.L.T?

A: No, but in Nigeria, Polytechnics shoulder the teaching of the course most.

Q: What are the credentials required for the admission of students to study S.L.T?

A: It is the same requirement as for other science courses i.e credit in at five subjects including mathematics and English Language at not more than two sittings in WAEC, NECO or GCE.

Q: How many years of study does SLT require?

A: It is only two years (4 semesters) in Polytechnics to bag the National Diploma as a Technician and another two years for HND (Higher National Diploma) which makes you a Technologist.

There is a break of one year for industrial training program before proceeding to HND. However, University requires straight four years duration.

Q: Where can I work as a Laboratory technician?

A: This has been discussed in this text.

Q: Can I study courses like Medicine, Pharmacy with a Diploma in S.L.T?

A: That has been allowed years back but it's no more allowed in our varsities. Studying science courses like chemistry, microbiology, botany, zoology and so on with the diploma is allowed using direct entry (DE) channel.

Q: What is the relationship between Medical Laboratory Scientist and Science Laboratory Technology?

A: Discussed in chapter one of this text.

NB: *ANY OTHER QUESTIONS CAN BE FORWARDED TO THE AUTHORS CONTACT*

CONCLUSION

Science Laboratory Technology is a course which fetches its students comfort especially those that are serious with it. As we might have noticed that the course is governmentally recognised. A nation is not yet complete without the science lab technician.

I herby implore the present, done and prospective students of SLT to pay good attention to the course as it is gold to be dug by us.

Great NASTES!!!

REFERENCES

www.nislt.gov.ng

www.prospect.co.uk

www.targetjob.co.uk

www.federalpolyilaro.edu.ng.